THE LANGUAGE OF THE CELL

THE McGRAW-HILL HORIZONS OF SCIENCE SERIES

The Science of Crystals, Françoise Balibar

Universal Constants in Physics, Gilles Cohen-Tannoudji

The Dawn of Meaning, Boris Cyrulnik

The Realm of Molecules, Raymond Daudel

The Power of Mathematics, Moshé Flato

The Gene Civilization, François Gros

Life in the Universe, Jean Heidmann

Our Changing Climate, Robert Kandel

The Chemistry of Life, Martin Olomucki

The Future of the Sun, Jean-Claude Pecker

How the Brain Evolved, Alain Prochiantz

Our Expanding Universe, Evry Schatzman

Earthquake Prediction, Haroun Tazieff

CLAUDE
KORDON

THE LANGUAGE OF THE CELL

McGraw-Hill, Inc.
New York St. Louis San Francisco Auckland Bogotá
Caracas Lisbon London Madrid Mexico
Milan Montreal New Delhi Paris
San Juan São Paulo Singapore
Sydney Tokyo Toronto

English Language Edition

Translated by William J. Gladstone
in collaboration with
The Language Service, Inc.
Poughkeepsie, New York

Typography by AB Typesetting
Poughkeepsie, New York

Library of Congress Cataloging-in-Publication Data

Kordon, Claude.
[*Le Langage des cellules*. English]
The Language of the Cell / Claude Kordon.
p. cm. — (The McGraw-Hill *HORIZONS OF SCIENCE* series)
Translation of: *Le Langage des cellules*.
Includes bibliographical references.
ISBN 0-07-035875-3
1. Cell interaction. 2. Cellular control mechanisms.
3. Cellular signal transduction. I. Title. II. Series.
QH604.2.K668 1993
574.87'6—dc20 92-13173

The original French language edition of this book
was published as *Le Langage des cellules*, copyright © 1991,
Questions de science series
Hachette, Paris, France.
Series editor, Dominique Lecourt

For Monique

TABLE OF CONTENTS

INTRODUCTION

What is the cell like? Is it cozy and intimate, like a small room? Austere like a monastery and grim like a prison? Or does it suggest, more often, the orderly bustling of bees in the compartments of a honeycomb? Before it came to refer to the common unit of all living organisms, this term, since its initial entry into the life sciences of the 17th century, has been invested with powerful intellectual, social, and emotional connotations that have affected the history of biology down to the present. Thus, it was applied to the well delineated individuality and even to the closed individualism of a hypothetical "atom" of living matter; but it has also been used in reference to the opposite notion of the cooperative activities of individuals who, like bees, are "entirely devoted to the commonwealth."

Georges Canguilhem has written definitive texts on these changes in meaning by placing them in the context of contemporary political events and prevailing social doctrines. In his book, Claude Kordon shows how today the cell is enmeshed in a new cluster of metaphors whose concrete value for scientific discovery is demonstrated, but whose undeniably suggestive imagery ("signals," "messages," "transmitters," "receptors") must constantly be anchored by precise biochemical content to keep the mind from going overboard into pure speculation. Such metaphors are just analogies: cells, of course, do not have a language in the sense that we humans do.

It could be be argued that the cell is not a mere term or abstract notion: it was first an observation; now its subtle structure is an object of study in the laboratory where thousands of researchers have been scrutinizing it for decades. True enough, but we must remember the conditions under which this observation of the cell became meaningful to science. Historians of biology agree that the term cell appeared for the first time in a work by Robert Hooke (1635–1703), published in 1665 and entitled *Micrographia or Some Physiological Descriptions of Minute Bodies Made by Magnifying Glasses*. Robert Hooke was an influential figure in a number of learned societies and contributed to numerous branches of science and technology.

In his *Micrographia*, he enthusiastically relates various and sundry observations made possible by the microscope, from the point of a pin to a razor's edge, from a piece of fabric to a grain of sand. Among those observations, he describes a thin slice of cork. He uses the word "cell" to designate the empty spaces of the plant tissue demarcated by "cellulose" walls, a name subsequently given to the principal structural element of plants constituting the rigid cell walls. Hooke wrote: "I was able to see with extreme sharpness that the fragment was entirely perforated and porous, very much resembling a honeycomb." Apparently, it is all there: the word and the metaphor.

Should Hooke then be credited with the discovery of the biological concept of the cell, as has been done so often? Certainly not: since he was not a naturalist, he had no interest whatsoever in the intimate structure of plants, and even less in seeking to discover the fundamental constituent unit of living beings. What caught his attention was the porosity of the materials he was studying. We must not yield to a retrospective illusion. The fact that Hooke used that particular word does not mean that along with it he gave us the modern, living concept of the cell. Contrary to a popular empirical view that still holds some appeal today, simple observation is not the first step in the acquisition of scientific knowledge; its role, which is surely essential, must be guided by a preexisting question.

This explains what many have found so puzzling, namely, that Hooke's "discovery" was not followed up. There was no trend of ideas where it could fit and be interpreted and analyzed as a scientific fact. Although the Italian anatomist Marcello Malpighi (1628–1694) and the English botanist Nehemiah Grew (1641–1712), both contemporaries of Hooke, did study the microscopic anatomy of plants and made similar observations, their interest was limited to the plant kingdom. And while they noted the presence of what we call cells, they considered them to be just one aspect of a plant's structure among others, albeit a singular one. Thus, the scientific career of the concept of the cell had not yet been launched.

It was not until the beginning of the 18th century that the question of a common composition for the plant and animal kingdoms was raised. Pierre-Louis Moreau de Maupertuis (1698–1759) took from the philosophy of Gottfried Wilhelm Leibniz (1646–1716) the conception of the monad (though he did not borrow this term) to explain the formation of organisms by the union of elementary molecules. His *Essay on the Formation of Organized Beings* was published in 1765. To each of the particles, he did not hesitate to assign principles of intelligence, desire, aversion, and memory. Among the cells, he detected "affinities." The term caught on, and is even used in the psychology of love. Today, we find it in biochemistry, but stripped of its psychological content. Buffon (1707–1788), a fervent disciple and translator of Isaac Newton (1643–1727) and author of a preface to Newton's *History of Fluxions*, sought to find in the living body the discontinuity of matter and to align the world of beings and the world of things, as seen through the mechanics operating within the same atomism; this is the purely speculative function of the notion of the organic molecule. His Newtonian philosophy, more rigorous than that of Maupertuis, explained the aggregation of molecules by a law of morphologic constancy that he paradoxically referred to as an "inner mold."

Though the question had been raised, the answer, as we see, was still pending: no connection was made between these supposed molecules or particles and the actual cells observed in plants through the microscope.

This pattern of thought ran into a major obstacle, namely the dominance, since the 16th century, in line with the ideas of Aristotle, of the notion of fiber which was always used when identifying the structural and functional unit of the living body. Descartes, despite his opposition to Aristotle, had supported the "fibrillar" theory in his *De l'homme* [Treatise on Man] (written in 1634).

At the same time that Buffon published the first volumes of his *Natural History*, Albrecht von Haller (1708–1777) gave a systematic and extensive version of this theory which met with enduring success in his famous *Elementa Physiologiae Corporis Humani* [Elements of the physiology of the human body]. He examined muscles, nerves, and tendons and showed that all the organs were composed of only one sort of fiber assembled in complex ways in various arrangements. And while he did mention the presence of "diamond-shaped" cells dividing the cavity of muscle fiber, the latter was the fundamental reality on which the crux of the arguments were based.

Thus, the formulation of the "cell theory" required a veritable "intellectual conversion" (G. Canguilhem), which did not come as a sudden event and was not the outcome of some revelation, as the term might suggest.

The groundwork was laid at the beginning of the 19th century by many scientific studies in botany, zoology, and medicine, and also by the daring views of a school of philosophy that later was disparaged more than it deserved, that of the "natural philosophers," the passionate center of German romanticism. The usage of the term *cell* first gained prevalence in botany. Quite naturally, it was a botanist who gave us the first formulation of the cell theory, though it was limited to the plant kingdom. In 1838, Matthias Jakob Schleiden (1804–1881) published an article that has been handed down to posterity, entitled "Contribution to Phytogenesis," in which he asserted that "every plant somewhat more highly organized [than an alga or a mushroom] is an aggregate of isolated, individualized, well-defined beings consisting of cells." A decisive step came a few years later when Theodor Schwann (1810–1882) extended this theory to all living organisms. But even before that step could be taken, a few preconceived concepts and even some prejudices had to be overcome in the analysis of the structure of animals. In the early 19th century, physiological research was fettered by the dogma that the ultimate unit of living matter consisted of "tissue." This dogma received some sort of scientific consecration in the considerable body of work of the great anatomist Xavier Bichat (1771–1802). What bestows properties on an organ is the nature of the tissue of which it is constituted. Appearances to the contrary, there is not much diversity in tissues: there are only as many as there are systems (nervous, vascular, skeletal, etc.). Bichat listed twenty-one in the human body (*Treatise on Membranes*, 1800). Thus, physiology was linked to anatomy.

Since tissue appeared to be the ultimate reality, there was no use in looking any further. Bichat was firmly against the use of the microscope—an astounding position, in retrospect: "When we view things in the dark, each of us sees them in a different way, depending on how they affect us."

For decades, the scientific authority of Bichat's work contributed to discrediting the use of the microscope to such an extent that Auguste Comte (1798–1857), an admirer of Bichat, gave a solemn philosophical justification to support that view in Lesson 41 of his *Course of Positive Philosophy*: "The abusive use of microscopic studies and the exaggerated faith still too often placed in such an equivocal method of investigation contribute first and foremost to lending some speciousness to this fantastic theory (the cell theory)." Just as in sociology, where the individual is an abstraction and the real social unit is the fabric of the family, the cells in biology—the "organic monads" as Comte calls them—are abstractions.

In order for the cell theory to become accepted, the microscope had to be revived over the objections of most of Comte's disciples, who were also eminent physicians and biologists. This was achieved mainly by

Henri Milne-Edwards (1800–1885), a French zoologist and physiologist of Belgian origin. But the tissue dogma also had to be destroyed, which required a complete philosophical about-face. Comte, the founder of positivism, greatly deplored this development from the very outset. The first decisive contribution to this turnaround was made by Lorenz Oken (1779–1851) in his treatise on *Generation* (1805): he drew a parallel between the body of large animals and that of microscopic creatures, and put forth the idea that complex living organisms were formed by the combination of simple living organisms through "fusion" into a unified whole that transcends them, rather than by the aggregation of primitive units (the much discussed "infusorians") into a composite whole.

The French physiologist and science historian Marc Klein has shown that this conception, based on the empirical results of "protistology" (the science of protozoans and other single-celled organisms), was inspired in its theory by the school of "natural philosophers" and particularly by Goethe and Schelling. In fact, the naturalist Theodor Schwann had as his teacher the great German physiologist Johannes Müller (1801–1858) who belonged to this school of thought. A straight line runs from the political philosophy of German romanticism (which emphasized the significance of the community to the detriment of individualism) to this physiological conception of the organism.

Even though Schleiden and particularly Schwann certainly introduced the modern concept of the cell as the basic unit of all living beings, the question of the genesis of cells remained very hazy in their writings, and they did not really venture into a description of their internal constitution. Schwann, in particular, took up the theory that cells were formed from a "structureless substance" that he referred to as the "cytoblastema." This was the theory of free formation, "spontaneous generation" of cells in an initially unorganized substance that would subsequently be given the name of "protoplasm," proposed by the Czech Purkinje in 1840 and used especially by the German botanist von Mohl (1805–1872).

In 1855, these speculative empirical views on the origin of cells were shattered by Rudolf Virchow (1821–1902). A few years later, in his *Cell Pathology*, he finally gave the answer in Latin in an aphorism that has become famous: "*omnis cellula a cellula*." Every cell comes from a preexisting cell (and not from a formless substance).

It has often been said that since then biology had found its atom, in the modern sense of the term. Like physics, and in large measure under its influence, contemporary biology would soon give up the idea of an indivisible element. Biologists would scrutinize its composition and discover its extreme and disconcerting complexity. François Jacob wrote: "With the cell theory, biology acquired a new foundation because the unit of the living world was no longer based on the essence of beings but rested instead on a community of materials, composition and reproduction. The particular quality of living matter requires that we look at its fine organization." In the second half of the 19th century, much of the effort of biologists was devoted to the cell. Cytology studied its inner space, and physiologists, embryologists, and theoreticians of heredity all focused their attention on it.

The historic year 1859 saw the publication of *Cell Pathology* and also of Charles Darwin's (1809–1882) *On the Origin of Species by Means of Natural Selection*. Today, we can say that the history of the life sciences was suspended for nearly a century because of the necessity of uniting these two great theories—the cell theory and the theory of evolution—which were developed quite independently of one another and followed completely different thought processes. This history did not follow a nice straight line as we sometimes imagine when we look at a simple chronology of the results.

Before these approaches could be united, physiology first had to experience its revolution. How were these units—the cells—organized to form an organism? In Paris, Claude Bernard (1813–1878) was a staunch supporter of the cell theory. In his *Lessons on the Phenomena of Life Common to Animals and Plants* (1878–1879) he gave new life to the politically sensitive metaphors, as he wrote: "The simplest form in which living matter can occur is the cell. The cell is already an organism: that organism may be a distinct being by itself; it may be the individual unit of a society whose aggregate is an animal or a plant."

The major concept of the *milieu intérieur* (internal environment), which creates a basis for the study of the physiological mechanisms whereby the organism adapts to the external environment and maintains its unity, provides the science of physiology with its own autonomous domain. But it also introduces a principle of coordinated action between the cells and the organs. This coordination is controlled by the major functions of living organisms: nutrition, locomotion, reproduction, etc. Through a remarkable conceptual reversal, the role of the organs and systems would now consist in combining the appropriate conditions for the life of the cells (temperature, humidity, and so forth); anatomy is subordinated to physiology.

With Louis Pasteur, the founder of microbiology (1822–1895) and a friend and colleague of Claude Bernard, chemistry completed its entrance into the life sciences. Pasteur's famous work on fermentation settled once and for all by means of experimentation the question of "spontaneous generation": living matter arises only from living matter, like from like. The same investigations provided a fresh look at organic substances: henceforth, their molecular structure, the position of the different atoms that make up the molecules, and their optical properties would be studied. All properties of substances, Pasteur asserted, are a function of three factors: "the nature, proportion, and arrangement of their constituent units."

The end of the century was marked by the rapid development of biological chemistry, which analyzed the cell content and studied the constituents of the cell and its reactions. The term biochemistry, introduced in Germany by Justus Liebig (1803–1873), then made its appearance. And it was discovered that living tissues are composed of three types of substances: sugars (saccharides), fats (lipids), and albuminoids (called "proteins" by G. J. Mudler in 1838). Chemists soon learned how to produce in the laboratory the reactions in which these molecules are involved, by using more sophisticated techniques of purification and analysis. They observed how slowly these reactions take place when reproduced under these conditions, at body temperature. They had to presume that some as yet undetected accelerating agent must exist. Indeed, they soon isolated for each reaction a particular "enzyme," namely, a molecule that increases the rate of the reaction tremendously by catalyzing it. Thanks to the discovery of enzymes, the tiny space of the cell was open to inspection by biochemists. Inside, they discovered a very dense and very complicated network of chemical operations that would remain a mystery until the middle of our own century.

Finally, we should add that a theory of heredity was indispensable if there was to be a productive connection between the theory of evolution and the cell theory. However, on this score, the Darwinian theory of natural selection had only taken up again certain ideas borrowed from the "generation" theories of the 18th century. As we know, the decisive importance of the work of Gregor Mendel (1822–1884) was not recognized until the early 20th century, with the contributions of Hugo de Vries, Karl Correns, Erich Tschermak and William Batcson. And still, their studies were only in the area of formal "genetics." It was not until 1944 that Oswald Avery identified DNA (discovered long before, in 1869, by Miescher who called it "nuclein") as the material forming the basis of heredity.

The picture was about to change completely when, in the wake of this discovery, the information theory burst onto the scene in the life sciences. One of the founders of quantum mechanics, Erwin Schrödinger (1887–1961) helped to proclaim this new movement in his famous book *What is Life?*, published in 1945.

Within a span of a few years, cytology, microbiology, and biochemistry would finally be coordinated under the command of the transformed genetics, now linked to Darwin's theory of evolution. Quite recently, as we shall see, a linkage was established between all these disciplines and embryology through the unexpected bypass of the neurosciences. This great and radical change, which has brought as much anxiety as enthusiasm to our lives, appears to be an indirect consequence of the powerful revolution that had transformed physics in the 1920s. That impetus imposed upon molecular biology a particular language and method of reasoning inspired by the communication and language theories that almost led to an impasse in the 1960s.

The reader will discover in these pages how such an approach, when cautiously and rigorously implemented, continues to inspire inventive thinking coupled with extremely refined experimental protocols. Claude Kordon initiates his readers to the "language of the cell" as he leads us on a breathtaking journey into the subtly intricate heart of living organisms, a world buzzing with messages of incredible precision. The cell appears as a tiny "chemical plant"; we cannot help but marvel at the monitoring and self-correcting devices for detecting and rectifying errors and at the strategies—and we might even say the "neat tricks"—that it uses for this purpose. You might react to this description by thinking of living matter as something remote and even inhuman. How wrong that would be! For we are part of it. We will see how its tiniest failures result in human catastrophes. And despite these painful and unavoidable examples, the lasting impression left in our minds will be the dynamism and enthusiasm of research: on the horizon is a new revolution in physiology as well as in pathology, with pharmacology and medicine ready to harvest its best products.

Dominique LECOURT

THE LANGUAGE OF THE CELL

I

THE CELL AND ITS SIGNALS

It has been known since the 19th century that all living things, whether animals or plants, are made up of cells. Back in the 17th century, when the English physicist Robert Hooke (1635–1703) looked at a thin slice of cork under the microscope, he discovered large numbers of small cells. But it was not until the cell theory was propounded by Rudolf Virchow (1821–1902) that the cell was identified as the basic structural and functional unit of living organisms.

Cells exist in a very great variety of sizes and shapes, as well as functions. Their average size is around one hundredth of a millimeter, though in some special cases, such as a hen's egg, it may reach a few centimeters. Of course, their number varies in different organisms: there are unicellular (single-celled) organisms and then there is the human body, which contains more than a thousand billion cells of various shapes—round, flat, elongated, star-shaped—depending on the tissue to which they belong.

Naturally, we can describe the cell by itself—its membrane, its nucleus, its liquid phase (the cytoplasm), and its framework, a true matrix (the cytoskeleton) consisting of semirigid filaments. Actually, the fulfilment of the cell's functions, and more generally the survival of each cell, are dependent on the cell's ability to adapt to its environment and hence on the coordinating signals it receives from it.

Every cell has the ability to detect signals from its environment. These signals, transmitted essentially in the form of chemical molecules that the cell has learned to recognize, are then decoded, that is, organized into messages to which the cell can respond based on its genetic repertory, for example by contraction in the case of a muscle cell or by hormonal secretion in the case of a gland. In this sense, we can say that there is a "cell language."

We can gain a basic idea of this language by examining a primitive animal composed of a single cell, the paramecium. It has already developed the ability to respond to biological or inorganic chemical molecules present in the liquid medium in which it lives. These substances act to attract or repel the paramecium through positive or negative tropism: thus, they can direct its movements by a gradient perception of signals (a *gradient* is a change in the value of a physical quantity in space as a function of its distance from a source).

Signals and messages are carried by minute particles of matter whose energy requirements are incredibly low. Thus, a photon striking a retinal cell will induce very slight reactions. When amplified by chemical processes, these reactions will eventually activate the entire receptor cell. This is one of the secrets of visual function that have been discovered at the cellular level.

Let us move up a notch in complexity and observe an insect. Like any other multicellular creature, it has specialized sensory cells able to decode the signals carrying information relating to its environment, and make an appropriate behavioral response. These signals may concern physical conditions in the environment such as temperature or the alternation of day and night; they can also supply more complex information, for example of a social nature. Such signals include pheromones, vectors of communication between different individuals, used in particular as sexual attractants, as items of information on the availability of a partner or the social hierarchy within a group.

The signals we are referring to consist of relatively simple chemical molecules derived from basic constituents of living matter: the amino acids, which are combinations of carbon, oxygen, hydrogen and nitrogen, with a molecular weight of around 100 daltons (100 times the mass of hydrogen). The simplest signals are themselves amino acid analogs (for example, glutamic acid or aspartic acid). In some cases, these amino acids are slightly transformed; this is the case with well-known transmitter substances (chemical mediators) such as epinephrine (adrenaline), dopamine, or histamine. But the most common signals (neuropeptides, polypeptide hormones) are made up of complex assemblies of amino acids, aligned in different sequences. There are also communication signals belonging to classes of chemicals other than proteins (sex hormones, adrenal hormones).

Enzymes, discovered at the turn of the century, were the first known example of the recognition of one biological molecule by another. Enzymes are very selective in catalyzing the chemical conversion of molecules (their *substrate*) that are present in the cells or around them, without being affected themselves by this reaction. In its basic form, the control of an organism over its environment hinges on this ability to convert molecules that are part of the environment. There is a "specific" relationship, meaning an exclusive one, between the enzyme and its substrate, which is explained by a special force of attraction (called *affinity*) of the two molecules. This concept of interaction between molecules can be generalized to subsume all aspects of cell communication of every molecule detected without being converted by the living cell. In this situation, the component that allows the detection of the signal molecule is called a *receptor.* The receptor also has an affinity for its signal; as in the case of enzymes, the interaction between the two components is governed by the law of mass action, a physical law that makes it possible to calculate the interactions between atoms on the basis of their respective properties.

LOCKS AND KEYS

We can consider the relationship between signals and receptors to be roughly equivalent to the relationship between "molds" and "impressions" or "locks" and "keys." Of course, like any metaphor, this one has pedagogic advantages as well as the potential for misunderstandings if taken literally.

Why do we speak of locks and keys with regard to cells and their relationships? Because the interactions between the molecules that diffuse freely around the cell and the corresponding receptors situated on the cell membrane that function like transducers use complementary tridimensional structures analogous to those between a lock and a key.

A biological molecule is defined not only by what we call its primary structure as determined by its empirical formula, but also by its spatial configuration. The three-dimensional structure of molecules depends on the way in which their basic units are assembled. Amino acids are very complex molecules; when added end to end, they form linear sequences, like a thread wound off a spool. But as it becomes longer, the thread becomes twisted and tangled like a ball of wool. In addition, bonds between special molecules all along the thread make the ball of wool rigid, somewhat like knots linking adjoining strands of the ball in several different places.

The shape of this ball—specifically its spatial, three-dimensional configuration—is not a matter of chance. It depends on extremely precise physical parameters defined by the geometry of the bonds between molecules and the distribution of their electrical charges. Therefore, complementary configurations are required in order for the signal and its receptor to interact. Thus, the receptor molecule exhibits sorts of cavities into which complementary outgrowths on the signal molecule can fit, somewhat like a push button fitting into its seat. The affinity of the two molecules is due to this complementarity of shape as well as of the spatial distribution of their electrical charges.

The affinity between the signal and its receptor is stronger as the complementarity of their configurations and their charges is more complete. That affinity defines at the same time an association constant, which reflects the probability that the key and lock will match, and a dissociation constant, which reflects the ability of the receptor to hold the signal (and which therefore determines the velocity of separation of the two units).

These time constants deserve attention. The gamut of affinity of the receptors is matched by a wide range of transmission velocities. The most affine communication, and hence the slowest, is not necessarily the most effective. Certain signals, for example, those of the epinephrine family, are specialized in the transmission of rapid information, with a critical time on the order of one thousandth of a second. Such a brief time corresponds to a weak affinity (and therefore with a strong dissociation constant). Conversely, other signals, such as neuropeptides and proteins, have much stronger affinities and their critical times of transmission are appreciably longer.

However, in order for a signal to accomplish its purpose, it must of course release its receptor after it has been received. If a signal were to persist indefinitely, the cell would be unable to receive the next message. Often, it would take too long to have to wait passively for this release to occur by natural dissociation of the signal-receptor complex. Therefore, cells have acquired the ability to regulate this dissociation themselves by causing the force of attraction that binds the two components of the complex to vary between a high affinity level ("unoccupied" state of the receptor) and a low affinity level (which favors the release of the "occupied" receptor). This evolution involves small changes in the configuration of the receptor. In other words, locks do not remain passive in relation to the keys.

We are starting to learn how to give a mathematical formulation to the concept of complementarity between signals and receptors. With this formulation, models can be created for the probable configurations of all the basic structural units by using computer processing. Only through such a procedure can we analyze the extremely complex molecular configurations which themselves are formed on the basis of the individual properties of each of their atoms (generally more than tens of thousands for a single receptor). In addition, the actual configuration of the receptor depends on its cell environment. The receptor is no mere appendage that the cell orients toward its external surface. It consists of long protein chains whose molecular weight is generally around 100,000 daltons and it may cross and recross the cell membrane three or five or seven times, depending on its nature and mode of action. Its mechanical interactions with the wall of the cell in which it is embedded as well as the interactions of its "floating" components (both within and outside the cell) with the physical parameters of the fluid environment may also influence its configuration.

Can the spatial configuration of the receptors be observed directly? The complexity of their structure makes it extremely difficult to do so, despite the real progress that has been achieved as a result of their direct visualization by means of advanced techniques based on crystallography and x-ray diffraction. However, if we just feed into a computer all the results of direct observation with the complete data on all the atoms concerned and their interferences (which appear as "knots" on the "ball of wool" of large proteins), a computer can calculate the probable configuration of a receptor. We can then try to construct artificial signal molecules matching the configurations thus obtained by calculation. The theoretical as well as the economic advantage of such an approach appears obvious. In the near future, pharmaceutical design and development will be completely different. Up to now, the design of new drugs has been based on the relatively old methods of traditional pharmacology developed when plants were our chief source of active ingredients. Those were empirical methods: the active principles were extracted, they were processed chemically or synthesized, and the most effective products were sorted out by means of laboratory tests. Soon, computers will make it possible to go beyond the known molecules whose activities have already been classified. Using the opposite approach, the theoretical optimal structure of a molecule will be predicted, based on the properties we would like it to have. In this way, we will be able to achieve the best interactions with a given receptor while keeping the side effects on other receptors to a minimum. Therefore, this will indeed be a new strategy for obtaining pharmaceutical products.

For the time being, this strategy is still theoretical in that it has not yet produced any major drug. But it has already enabled us to obtain some valuable variants of known drugs, and above all to understand the reasons why a drug (be it an enzyme or an antibody) is or is not effective. Similarly, using this new approach, we have been able to produce a number of novel compounds that can inhibit enzymes in our body. For example, this method led to the development of a very important enzyme for the cardiovascular system (the angiotensin converting enzyme or ACE). It converts the precursor of a chemical messenger (renin) to a more active signal (angiotensin). Excessive activity of this enzyme results in severe hypertension, which can be controlled by the new inhibitors. Several years ago, a related family of enzymes, the enkephalinases, was also obtained by this method. These enzymes degrade some of our "natural morphines," the enkephalins, one of the forms of endogenous opioids that modulate, in particular, our perception of pain.

The action of opium derivatives has been known for a long time. Opium is a narcotic extracted from the poppy (*Papaver somniferum*), whose medical uses in the West can be traced back to ancient times. The poppy is referred to as the "plant of joy" on a Sumerian tablet. In the 5th century B.C., after the Egyptians praised the "sleep-inducing" properties of the plant, Hippocrates and Theophrastus, the founders of Greek medicine, had already extracted from it a medication to relieve pain. When the mechanism of action of opium derivatives was investigated, it was found that they were capable of binding to certain tissues and to certain structures of the brain. Cells throughout the body were equipped with locks adapted to these signals. Obviously, these locks had not been selected through evolution so that they could be opened by "opium," "morphine," or "heroin" keys—keys for plant products which, except under highly exceptional circumstances, have no reason to circulate in the human body. But starting from these artificial signal molecules, scientists were able to trace back to the natural signal molecules for these receptors and discover a large family of mediators, the opioids, produced by the body itself; that family includes the enkephalins and endorphins.

The enkephalins bound to the receptor must subsequently be salted out so that the receptor can be ready to receive the next signal. However, to prevent the released molecules from interfering with the reception of the new signal, they also have to be destroyed. Enzymes (specifically, enkephalinases) associated with the receptor have the task of inactivating the old signal.

We know the broad medical and social implications of the problem related to habit-forming drugs, particularly in the case of opiate dependency. The development of "withdrawal" strategies is very difficult; therefore, researchers wondered whether prolonging the action of the endogenous opioids by inhibiting their breakdown might not be a good alternative to the temporary administration of substitute narcotics such as methadone, currently used with only relative success. The hope is that by acting on the natural opioids, the functions of these substances that have been disturbed by drugs can be restored more gradually and perhaps in a less traumatic way than with the current methods. Computer-designed enkephalin inhibitors are now undergoing clinical studies. The future will tell whether they are more effective than morphine substitutes.

SIGNAL TRANSMISSION AND THE CELL MESSENGER SYSTEM

The study of cellular communication is not limited to the modalities of signal reception; it also has to be concerned with the mechanisms involved in their transmission. Each cell acts as a communications unit, a miniature production facility that can manufacture and manage not only its supply of signals but also its raw materials and tools, such as the enzymes from which the signals are produced. Thus, the cell has channels for importing raw materials (essentially amino acids), and it also has a production line for the transformation of these units of material (protein synthesis or enzyme reactions). Simple signal molecules such as epinephrine, serotonin, or histamine are amino acids modified by the addition of a molecule containing hydrogen and oxygen or the elimination of a carbon-oxygen group. These transformations use enzyme mechanisms involving enzymes located in specialized structures (secretion granules or cell vesicles).

In addition, the cell can control the amount of its reserves of chemical mediators (transmitters) as well as the degree of intensity of their transmission. It can also dispatch the products that it manufactures to many different destinations: the membranes or cellular organelles in the case of its own component units (the "spare parts" required for its maintenance); the lysosomes where the excess or defective constituents are dumped; or the vesicles through which the cell exports its products.

As we can see, this is a complex and delicate system: the cell is the theater of operations for a whole array of molecules involved in movements of exocytosis and endocytosis to effect the transport of vesicles to the surface of the cell, the reuptake of signals or receptors by the membrane in some cases, their recycling within the cell, etc. To avoid any mistakes as to where all these products are to be sent, the cell has what amounts to a parcel delivery service that is guided by "addresses," by chemical "tags" or labels. Those labels generally consist of fairly simple molecules (often sugars) attached to the product being forwarded and recognized by the structure for which it is intended.

Certain "polar" cells—morphologically and functionally asymmetrical—use similar mechanisms to operate as carriers of the products that are directed from one side to the other. In this manner, several constituents (such as hormones, antibodies) collected in the blood by the exocrine cells of the maternal mammary gland are labeled, transported and finally incorporated in the mother's milk. Thus, the infant benefits from part of the immunity acquired by the mother.

Luckily for us, the cell messenger system does not get out of order very often; when that does occur, it is generally due to genetic defects. For example, a mutation can mask the anchorage site of a "label" and prevent it from binding to the molecule being transported. There are known cases of perfectly normal hormones which are unable to be conveyed properly to the granules and therefore accumulate in the cell and cannot be secreted. Thus, a genetic defect observed in the rat affects its ability to secrete vasopressin, a pituitary hormone that regulates blood pressure and the reabsorption of water by the kidney. It was long believed that this abnormality was due to the pathological absence of the type of cell responsible for secreting the hormone. We now know that the cell is present and normal, but that a mutation affecting a single nucleotide—a single link in the chain of a gene that has tens of thousands of such units—produces an abnormality in the molecular sequence that controls the intracellular transport of the hormone.

The low-density lipoprotein receptor offers another example of such genetic errors. It is a somewhat special model of a receptor because its main function is to internalize and then sequester its signal, which in this case is a form of cholesterol. Many mutations of the gene coding for this receptor have been studied, and some of them produce recognition errors. When this occurs, cholesterol can no longer be reincorporated in the cell; it accumulates in the vessels, causing early atherosclerosis as a result of hypercholesterolemia. This is not hypercholesterolemia related to diet and due to excess cholesterol, but a genetic defect that impairs the use of normal stores of cholesterol by the cell. We are now able to characterize the mutations of those receptors and thus confirm the diagnosis of the disease; for this purpose, we just need a small biopsy specimen of tissue from an organ that expresses the receptor.

In diabetes, as in metabolic diseases involving lipoproteins, we have probes through which the molecular errors responsible for the disorder can be localized with great precision. In this way, we can distinguish among several types of diabetes that have different causes and therefore do not call for the same treatment. Thus far, however, we are able to correct only the consequences of these genetic defects; correcting their cause is, of course, far more difficult and would require modifying the pathological genes by means of interventions on the genome. This raises questions of ethics that go far beyond a scientific debate.

The cell also has monitoring units to control its own communications. Many cells thus have "auto-receptors," that is, receptors sensitive to the signals which the cell itself has produced and which tell it how many signals it has sent out around itself. When these autoreceptors are inhibited, the cell is generally carried away and transmits an excessive number of signals.

Finally, as mentioned earlier, the cell can inactivate these signals itself; it can also recapture them by mechanisms that ultimately "sequester" the signals already used in order to recycle them for possible reuse at a later time.

These extraordinarily meticulous processes have been studied with new tools that make direct observation possible and generally have been devised by immunologists and cell biologists working together.

Specific antibodies that will recognize signals or their receptors can now be manufactured artificially. Using what are known as cytoimmunochemical techniques, those signals can be not only measured but also visualized to determine their cellular or intracellular localization. The principle employed consists in tagging the signal with an antibody which itself has an artificial label, such as a product that produces a color reaction, or with a radioactive molecule. By visualizing the locations where the signals and receptors are expressed, their production sites and distribution in the body can be mapped.

Using similar methods based on a new generation of markers, it is now possible to locate and measure accumulations of signals within the human brain itself. Such noninvasive methods enable us to pinpoint, from outside the body, electromagnetic radiations or high-energy particle flows emitted by collisions between matter and antimatter (electrons and positrons) and occurring at locations where the antimatter marker is concentrated.

ADHESION AND COOPERATIVE SIGNALING

Up to this point, we have been looking at fixed receptors and moving signals. But the same (or nearly the same) rules may apply to fixed signals. Interactions among fixed signals on the surface of cell membranes can lead to physical contacts between cells that express complementary keys and locks at their surface. Such contacts play an important role in the immune system. Two essential components of that system, lymphocytes and macrophages, can adhere together temporarily and mutually present to each other an antigen, usually a foreign molecule (elements that may be of microbial origin or may have entered the body accidentally). The foreign body will then be destroyed or inactivated by "antibodies" produced against it by the lymphocytes. The presentation procedure is essential to make sure that the response to the antigen is appropriate and to coordinate the action of the different immune system partners that join forces against foreign intruders.

Those reactions essential to the body's defense are subjected to strict control. For the researcher, the strategies of this control represent highly interesting models, to which we shall return again later. Their purpose is first of all to identify, among all the molecules of the "self" (blood cells, hormones, signals, elements used in the formation of organs that circulate in the bloodstream and the extracellular fluids, etc.), any molecules of "nonself" that might be concealed. The soldiers of the immune system, and particularly the circulating cells called lymphocytes and the macrophages, have developed multiple safeguard mechanisms to eliminate "inappropriate" reactions and avoid confusions between self and nonself molecules. Such confusions would result in serious disturbances characterized by the onset of autoimmune diseases. In this type of disorder, the defense of the self paradoxically turns against the self, and cells or molecules belonging to us are destroyed by our own immune system. For example, there are diseases of the thyroid that are caused by a gradual destruction of the cells of this gland or by the masking of its receptors by antibodies. Also, a type of diabetes attributable to the inactivation of the insulin receptor by antibodies has recently been discovered. Such errors can also be caused when molecules that are normally concealed enter the circulation by mistake, for example following a lesion which exposes the internal cell constituents. Not being accustomed to encounter this molecule in a free state, the body's defenses "immunize" against it and eventually destroy it, even in healthy cells.

Autoimmune diseases can also be triggered after a normal defense reaction, followed by a confusion in the recognition of the antigens involved. For instance, bacteria and viruses produce molecules that present great similarities to molecules of the self. As a result, the antibodies directed against these bacteria are not always able to make the necessary distinction and will damage parts of their own body.

Furthermore, immune system defense requires a learning process. The lymphocyte system retains in its memory nonself substances that it has seen once in its life, even if the substances are no longer present. Thus, it can mobilize very quickly as soon as it finds them; this is the principle on which vaccination strategies are based. But, of course, a body that has "learned" to confuse certain of its molecules with foreign molecules also retains the identification in its memory and perpetuates behavior that is suicidal for its own cells.

We can see the importance of the mechanisms which precede the activation of the body's defenses in controlling the appropriateness of the immune response. This control involves a "dialogue" among several cellular elements after they have compared their assessments of the danger of invasion by foreign cells or antigens. This is the purpose of the presentation of an antigen by a macrophage to a lymphocyte, which is in turn responsible for activating another lymphocyte that produces antibody. As we have mentioned, the physical contact, the temporary adhesion necessary for this presentation, is provided by complementary locks and keys which are attached to the surface of immune system cells and are called cell adhesion molecules (CAMs). A common characteristic of nearly all these molecules is that they contain a short sequence of four amino acids, namely, arginine (R), glycine (G), asparagine (N) and serine (S), expressed in one direction in the "key" molecule and in the opposite direction in the "lock" molecule (RGNS/SNGR), so that each sequence is a mirror image of the other.

This arrangement is not limited to CAMs but also occurs in a large number of vital processes where controlled physical contact between cells appears essential. The formation and the positioning of the organs during their development is accompanied by a very large number of cell migrations. Of course, these migrations must take place in a strict order, follow accurately marked paths, and come to a halt when they reach carefully located targets. These mirror sequences serve to locate the "road signs" and "beacons."

During the development of the body systems, few migrations attain the degree of complexity that they display in the nervous system. In neuronal migration, the growth of nerve cell projections (dendrites and axons), their juncture with their targets (namely with other neurons that are sometimes situated at a considerable distance), are controlled by remarkably coordinated signaling mechanisms (road signs and beacons) based on the same principles and the same chemical sequences as the lymphocyte adhesion molecules. Derangement of these mechanisms, which fortunately is an uncommon occurrence, results in abnormal embryogenesis and may lead to the development of malformed fetuses called monsters, whose study is called teratology.

However, foreign organisms such as parasites, bacteria or even viruses can also use the code for these mirror-image sequences to bind to the cells they infect or in order to choose their destination in the cell by using molecular labels addressed to the cell, that "fool" the communication system of their host.

These communication perversions by means of decoy signals lead to a race between the host and the parasite which, at the risk of disappearing, must recognize its target cell and penetrate it before it is recognized by the body's defenses. Because of the importance of these common sets of locks and keys in infectious processes, attempts have been made to "fool" the parasite, for example by flooding the host organism with synthetic molecules that carry the famous coding sequence with the ubiquitous four amino acids of CAMs mentioned above. The purpose of that strategy was in effect to block access to all locks by saturating them with false keys. Although this method may be successful in slowing the process of infection, it cannot yet be used as a preventive treatment: when locks are rendered less accessible to the parasite as well as to normal communication within the host, far too many adverse reactions are produced.

Not the least paradoxical aspect of the host-parasite relationship is that animal species have selected the same communication signals even though they are very far apart on the evolutionary scale, having appeared on earth separated by an interval of hundreds of millions of years.

II

DECODING AND THE CELL MESSENGER SYSTEM

We have just seen how the cell developed the ability to transmit and recognize signals. Until recently, it was believed that this provided a sufficient explanation of cell regulation. As a first approximation, we could be satisfied with the explanation that chemical signals transmit stimulatory or inhibitory impulses to each cell to control its level of activity through orders to "accelerate" or "inhibit" its activity. The development of numerous drugs designed to act as artificial cell accelerators or inhibitors has adopted this approach.

Recent discoveries, however, have shown that cell-to-cell communication is not that simple. The situation can be summarized by three observations: first, each cell is not simply controlled by an accelerator and an inhibitor; actually, it has the ability to recognize a great many different signals. Secondly, the number of signals discovered in the body has increased tremendously: while barely two dozen were known twenty years ago, hundreds have now been identified. Lastly, signals have become universally commonplace; they are not, as was formerly believed, characteristic of an organ or function ("nervous system mediators" or "immune system mediators") but are all found, regardless of their chemical makeup, in nearly all organs and associated with nearly all functions.

Let us take a closer look at the observations that led to these three discoveries. The finding that a cell can recognize many signals is the result of a better understanding of the key molecules and of the miniaturization of techniques to identify the locks of each cell. Thus, researchers have been able to complete the inventory of locks that characterize each type of cell, and to determine for example that the cell responsible for the secretion of a pituitary hormone, namely, prolactin, can respond to at least twenty different signals.

The discovery of new communication molecules is the result of the development of extremely effective chemical methods that make it possible to purify them and elucidate their structure, even if only a tiny amount of molecular material can be obtained from them. The total mass of certain signals in the body is hardly more than a billionth of a gram, and it has to be separated from myriads of other molecules present in the organs.

The older techniques permitted in the main relatively simple molecules to be detected, such as the classic mediators we have already mentioned—epinephrine, histamine, and acetylcholine. More recently, convergent evidence has suggested that there was also a more complex chemical category of signal molecules, the neuropeptides. Therefore, considerable efforts have been devoted to characterizing those substances, and it was promptly discovered that they were part of an extremely large chemical family. The researchers in charge of two groups in the U.S., Roger Guillemin, a Frenchman, and Andrew Schally, an American, received the Nobel Prize in 1977 for finding the formula of the first three neuropeptides (TRH, GnRH and somatostatin, all three of which act on pituitary secretions). But these discoveries did not come easily. After indirect evidence of the existence of these molecules had been collected, it took another fifteen years of work, of trial and error and controversy—not to mention considerable investments of tens of millions of dollars—before their exact chemical structure was discovered. This adventure is described in a book by Nicholas Wade, *The Nobel Race*, which reads almost like a spy novel; while seeking to identify the peptides involved, each protagonist was also trying to throw his competitors off the scent.

Ultimately, the structural formula of the three signals proved to be relatively simple, the smallest one consisting of only three amino acids! The techniques required to achieve these results were developed as the work progressed. Today, they have become almost routine, and the same work could be accomplished much more rapidly.

NEW FAMILIES OF SIGNALS

Aside from the advances in protein chemistry, we also have much more effective strategies for identifying unknown signals. Instead of isolating and identifying them one by one, genetic engineering methods make it possible to search banks of genes expressed by each organ for the genes encoding signals or for messengers that will translate those genes into amino acid assemblies. Once the sequence of the nucleic acids (DNA or RNA, easier to analyze than the sequence of the amino acids themselves) has been determined, the structure of the signals can be deduced. With this procedure, which can also be applied to the discovery of new receptors, new molecules can be described before even knowing the purpose that they serve. Using these methods, some laboratories have been able to characterize several signals or several receptors within a year. Thus, new signals are coming out nearly every day!

Since these methods were developed, it has been established that cells had a much wider array of locks and keys than was previously thought. When the list was prepared, it was found that these locks and keys fell into large families; to be sure, the signals and receptors of the same family differ slightly from each other, but they do have close structural and functional similarities. The existence of these families probably plays a very important role in the adaptation of living beings to their environment and therefore in their ability to survive evolution. The availability of several neighboring signals transmitting similar but not quite identical messages introduces a whole set of subtle variations in cell communication. These modulations add considerably to the available repertoire for communication among cells, by varying the affinity of the signals, the speed of transmission of messages or the type of receptor coupling. Those abilities are vital because the adaptability and ultimate survival of organisms depend on the diversity and sensitivity of their communication strategies.

Matching this extraordinary diversification of signals is a diversification in the ways they are used. Several signals from the same family can be transmitted together; some of them are not very stable and will be recognized only in the immediate vicinity of their source, whereas others can carry their information over great distances.

Some will transmit to different groups of cells messages that are designed specifically for each of their targets. For example, signals from the growth hormone family can deliver a lipolysis command (an order to transform and use fatty acids) to fat cells, while other molecules in the family will act on the liver or the pancreas to increase the availability of sugars. These combined actions will promote a coordinated mobilization of energy resources. In certain cases, the coordinated production of these specialized signals can even be done *à la carte*, that is, on demand from the target cells for which the information is intended. The different forms of the growth hormone, all derived from the same precursor, translated from a single gene, can be modified in this way by enzymes situated in the vicinity of the receptors themselves. Different enzymes, associated with the various classes of the hormone receptor, transform the precursor into the form best adapted to each particular receptor.

Awareness of this profusion of families, each of which consists of many signals, resulted in exploding the conventional concept of a mediator. Formerly, a signal had to fulfill some very restrictive criteria in order to qualify as a mediator. The mediators of the nervous system, such as the neurotransmitters, belonged to a very exclusive club; before a new signal was admitted, proof had to be adduced that it was synthesized in a nerve cell and released by it, that its original synapse contained the corresponding receptor, and that the enzymes locally present were capable of inactivating it once it had completed its mission. Those rules have had to be revised. In the case of most of the hundreds of newly discovered mediators, we still do not know whether they are actually recognized in the synapse or outside of it, and therefore whether gaps in the inactivation system might not enable them to diffuse much further than would be anticipated by the theory of synaptic transmission. We do not even always know whether their transmission and receptor sites really coincide. And yet, we are almost certain that they are authentic neuronal signals. Therefore, the club of neuromediators has become more democratic and will accept nearly all candidates, since it is not able to define new conditions for membership.

Let us stop for just a moment longer at the third observation: the universality of cell communication implies that no signals are attached exclusively to one organ or one function. For example, neuropeptides such as GnRH, TRH, and somatostatin, which we mentioned earlier, were first discovered in the hypothalamus, a special structure in the brain that controls autonomic functions (namely, regulatory functions, generally unconscious and involuntary, such as hormonal secretion, digestion, regulation of the heart's action and, at least to some extent, respiration). It was originally assumed, therefore, that the neuropeptides were specific signals for those functions. But it quickly became apparent that these same peptides occurred in other parts of the brain and even in many other organs, such as the digestive system, whose organization is of course quite different from that of the brain, or even in the immune system. Conversely, interleukins—communication signals that were originally "reserved" for lymphocytes—are also produced and recognized by the nervous system.

Thus, it seems that all organs potentially have access to the same stock of signal molecules, to the same set of communication tools, we might say. The well-known expression used by François Jacob, who some years ago spoke of "tinkering" in connection with evolution, takes on new meaning here. Selection pressure, he suggested, sorts out gene expression according to relatively empirical strategies, based on the relative usefulness or adaptability of their products. We can extend this concept to signal molecules: some of them have proved to be multi-purpose, usable under very varied conditions. They can carry out all sorts of assignments: provide local cell-to-cell communications and also long-distance communications. The best example is epinephrine, which acts both as a nervous system mediator and as a hormone.

Other signals, on the contrary, remain highly specialized and hence are of more limited use. Some, which are less common and are designed for very specific tasks, have a less ubiquitous distribution. This is the case with GnRH, a small peptide of ten amino acids, which is involved mainly in reproductive functions by contributing to the regulation of sexual behavior, reproductive hormones and external genital organs. On the whole, in many organs, the tinkered strategies of evolution have managed to stabilize signal tools with very general uses while maintaining the use of highly specialized signals in a few organs only.

The greater or lesser specialization of signals often hinges on their sensitivity to degradation mechanisms and therefore on their ability to transport information at a distance. Certain signals are specialized in long-distance communication. This is the case of hormones that circulate freely in the bloodstream and can act on any organ, provided of course that the latter has the right receptors. Other signals are limited to local communication, since their chemical properties do not permit them to be transported to a distant point without being destroyed. They are often prominent in "autocrine" or "paracrine" communication in which the signal acts on the very cell that transmitted it or on cells situated in the immediate vicinity. Aside from these differences in range of action, the traditional distinction between mediators of the nervous system, the immune system, or the hormonal system is no longer very pertinent.

We should also mention that the new signal detection methods have not only made it possible to find signals in nearly all tissues, but they have also shown that organisms differing greatly in terms of the time at which they appeared on Earth or of their phylogenetic status were capable of expressing the same signals, as we saw earlier in connection with the cell adhesion molecules.

CODED MESSAGES AND DECODERS

In light of the growing number of known signals, the fact that cells are able to recognize many of them at the same time, and finally the observation that they are universal to all organs and even to all organisms, we must revise profoundly the way we see cell communication. The problem is not so much to find out how cells address messages to each other as it is to understand how each cell manages to extract an identifiable message from the immense background noise of the thousands of signals that circulate around it. In other words, how can it make some sense out of all that cacophony? The problem that therefore arises is one of "decoding."

The cells' decoding mechanisms are located downstream from the receptors. They are based on complex chemical reactions that take place in the cell membrane and control all the responses of the cell to the messages it receives.

To understand the rules of decoding, we must first hark back to the locks referred to above and find out how receptors are organized in the membrane. Those receptors all have a recognition component (the complementary configuration to that of the signal key) and an "effector" component that will pass on the reception of the signal to the intracellular machinery. Effectors operate in two principal ways: they can either produce a very specific chemical modification of certain intracellular components (the addition of a phosphate group to a protein, a reaction called phosphorylation), or modify the electrical properties of the membrane by opening or closing channels through which ions (for example, sodium, potassium, chlorine or calcium) enter or exit the cell. The effector sites of receptors that act by phosphorylation are characterized by an enzymatic activity that produces the reaction. Those that control the opening or closing of channels do so through pore-like protein sequences that pass through the membrane.

There is a very large number of different phosphorylating enzymes, distinguished by the nature of the proteins that they are able to phosphorylate. There is also a very large variety of ionic channels that are distinguished by both the way in which they function and the type of ion that flows through them. But the interaction of a signal with a receptor always activates one or the other of these two classes of effectors.

In addition, effectors are characteristic of their receptor. This means that, as a rule, a given receptor always acts through the same type of enzyme or the same type of channel. Therefore, coupling to a given effector is just as much a part of the intrinsic properties of the receptor as its signal recognition properties. This law seems to hold regardless of the family to which the rcccptor bclongs, and also irrcspcctivc of thc organ or even the animal species that expresses it. In this sense, the rules for coupling appear universal.

Independently of their manner of coupling, however, there are two principal categories of receptors that differ by the way in which their recognition and their effector components hinge on each other. On the first category of receptors, the simplest ones, the two components are combined within a single molecular entity. By contrast, receptors belonging to the second category have several independent molecules.

For example, the insulin receptor or the growth factor receptors (which are important signals for the organogenetic programs) express a recognition sequence and a phosphorylating enzyme sequence on the same protein, in what are called different "domains" of its amino acid alignment. A slightly different type, the nicotine receptor (so called because nicotine has a signal-mimicking action upon it) for a mediator that has been known for a long time, acetylcholine, expresses its recognition sequence and its effector sequence—through a sodium channel, in this case—on different proteins; but these proteins are reassembled by covalent bonds to form a single molecular entity with a pentameric structure (five proteins linked together), as was demonstrated by the research group headed by Jean-Pierre Changeux at the Pasteur Institute.

As they became more complex, the receptors belonging to the second category acquired an additional regulatory mechanism: "coupling" molecules are intercalated between the receptor component and the effector component. These molecules are part of the family of G proteins (guanine-nucleotide binding proteins, to which guanylate triphosphate, or GTP, binds) that includes several types.

The advantage of "coupled" receptors is that they are more versatile as regulatory units: the coupling components can act as regulators of the level of signal amplification, or, to borrow a metaphor from electronics, of the level of "gain." Thus, the cell acquires better control over its own sensitivity. The "amplifier" (which can also act in the opposite way as an attenuator) operates in the same manner as does signal-receptor recognition. The receptor and coupling components come into contact on the basis of their complementary configurations; this encounter results in a modification of their configuration, somewhat in the manner of a "cascade" or domino effect, which propagates a deformation wave within the membrane. This deformation wave eventually reaches the phosphorylating enzymes (by way of small molecules known as second messengers) or the ion channels that are the effectors for the particular receptor. Once the reaction is over, all of the chain's components resume their original forms and return to a resting state.

Coupling proteins function somewhat like a railroad signal box: they help to select the transmission track, to choose the effector corresponding to each receptor. But these coupling chains are also "read heads" for messages received by the cell and are at the heart of their decoding mechanisms.

We mentioned earlier that the activation of certain receptors produces a phosphorylation reaction. In some cases, the substrate of this reaction is the receptor itself. Activation of the receptor results in a sort of "autotransplantation" of a phosphate radical. In that case, the binding of the signal modifies the properties of its own receptor so as to produce an increase or a decrease in its sensitivity. Such modifications are responsible for most of the "hypersensitivity" or "desensitization" phenomena of receptors. They play an important role in the exchange of information between cells, but they also account for certain aspects of acquired tolerance. A cell whose receptors have been desensitized by an excess of spurious signals, as under the influence of drugs, will cease to recognize its natural signals for a long time and thus become "deaf" to its environment. Obviously, its function will be perturbed for a lengthy period.

Phosphorylation, however, can also affect a different receptor from the one that triggered it. This is known as heterologous phosphorylation of one receptor by another. This property is responsible for an important characteristic of the decoding mechanism. In this instance, it may happen that a receptor will not be able to perceive its signal unless it has first been primed by the activation of another receptor. Let us arbitrarily designate these two receptors as *x* and *y*: the sequence *xy* will be recognized by the cell, while the sequence *yx* will not, since *y* can only be read after *x*.

Those rules constitute what we could regard as a cell language "grammar." Just as with any other language, meaning is not limited to the sum of symbols that express it; it depends on the way in which these symbols are combined and interpreted. The message *xy* is more than just the sum of signals *x* and *y*.

Here we find another manifestation of the diversity of cell language. Although the rules for coupling to receptors x and y remain universal, the same cannot be said of the grammar of their assembly, which is peculiar to each type of cell and represents the cell's "latent program." Thus, different cells appear capable of interpreting the same message in different ways. Two pituitary cells, each of which secretes its own hormone—prolactin and growth hormone—are stimulated in the same manner by the neuropeptides TRH and vasoactive intestinal peptide that control their activity. But this occurs only when they receive the neuropeptides separately; if they are given at the same time, the prolactin cell is greatly stimulated while the growth hormone cell ceases to respond. So, they both contain the same basic signals but do not have the same reading program. To put it briefly, they can decipher different "languages" using the same "alphabet."

The brain's communication capacity is enhanced by the fact that the billions of neurons of which it is composed have different "grammars." These deciphering rules also play a very important role in redundant (fail-safe) mechanisms of the immune system.

As was already pointed out, recognition errors by the mechanisms in the immune system can have grave consequences: unintentional destruction of elements of the self, failure to be alert to nonself antigens. The cells responsible for the body's defenses have selected protection systems that use different decoding combinations. For instance, a foreign antigen is perceived as foreign only if it is presented to the lymphocyte by another cell. But this condition is still not sufficient: in addition, the appropriate signals from the family of interleukins have to "confirm" the order to respond to the intruding antigen by the secretion of antibodies. Such confirmation is based on the previously mentioned rules of sequential signal presentation: binding of the interleukin to its receptor following exposure to the antigen permits the response to occur; if the order is reversed, there will be no response.

The sequential coding of information is a very general property of cell communication. We now have many examples of this, among which we can cite the two principal hormones that control reproduction in females: estradiol and progesterone. These two signals nearly always act in concert, but their effects on sexual behavior or on the development of the fertilized ovum in the uterus depend entirely on the order in which they reach their target tissues.

III

THE LANGUAGE OF INTEGRATED SYSTEMS

STRESS AND EMERGENCY SIGNALS

So far, we have only discussed cell-to-cell signaling. How are the exchanges of signals integrated in the general coordination plan of organs? Let us take the case of the interface between the brain and the hormonal system, a field of study called neuroendocrinology. This has traditionally been a well-studied discipline in France. It was a Frenchman, Jacques Benoit, professor at the Collège de France who—at the same time as the British neuroendocrinologist Geoffrey Harris, yet independently of him—described for the first time in the early 1950s the interaction between specialized neurons situated in the hypothalamus at the base of the brain and the pituitary, an endocrine gland, often referred to as the master gland. The relationship between these two structures contributes to the maintenance of hormonal homeostasis and to the adaptation of the body's autonomic functions to changes in its environment through hormones that act as regulators of many organs. A cooling of the external temperature perceived by the nervous system's sensory organs triggers an increase in the production of heat and at the same time a decrease in its dissipation. Both responses involve concomitantly nervous system mechanisms (changes in behavior, chilling, etc.) and a hormonal link: stimulation of the thyroid. The concerted action of these two coordinated mechanisms thus functions like a thermostat.

Involved here is a neuroendocrine reflex of peripheral origin (the sensory organ), with secretion of a hormone as the response. There are many neuroendocrine reflexes of this type, not all of them related to the adaptation of organisms to their environment. For example, suction on the nipple during breast-feeding produces the release of several hormones by way of a reflex action, particularly prolactin which plays an important role in the production of milk and oxytocin, which ejects the milk and facilitates feeding by the infant. This reflex is conveyed by sensory pathways to the brain which relays the information to its neurons interfacing with the hormonal system.

The organism's adaptive reactions are not necessarily physical. They must also allow individuals to adapt to their social environments and therefore include behavioral responses such as flight or fight or the regulatory mechanisms of social hierarchies. When an animal is in an emergency situation or in distress, it undergoes hormonal changes that tend to facilitate its behavioral response; these modifications reflect choices of priorities and are intended to adapt the body's energy economy to the flight or fight strategies that will be implemented. Emergency signals will therefore override the organism's regular communications, just as an ambulance with its siren going has the right of way over the regular system of traffic signals. The temporary suspension of regular communications between cells to make way for urgent signals is part of what we call stress reactions, which were described for the first time in the late 1940s by Hans Selye, a Hungarian-born Canadian researcher. Turning his interest to the mechanisms of stress, Selye discovered that the response to aggression is part of a much broader set of strategies whose purpose is to allow the organism to adapt to the challenges of its environment. Toward the end of his life, Selye defined stress even more generally as "the response of the organism to any demand made upon it."

Let us take the example of an individual thoroughly trained to follow instructions, such as an air traffic controller or any other professional, to confront unpredictable events with complete control. This individual's reactions to each of these events will be accompanied by specific hormonal secretions. The blood levels of vasopressin and oxytocin will change. These two hormones produced by the brain are involved, among other situations, in mechanisms of learning, including the acquisition of conditioned responses to new instructions. Along with this, the adrenal gland releases greater quantities of a signal we mentioned before, norepinephrine (noradrenaline). This secretion accompanying the adaptive response may be compared to control lights indicating that certain instruments are in operation. But when the operation does not achieve the expected result, the organism is confronted by failure, which may kindle a distress reaction.

The distress reaction will also be accompanied by episodes of hormonal secretion, but this time involving the liberation of other adrenal hormones as well as pituitary hormones, including a morphine-like substance, β-endorphin. To use the same metaphor as above, the body signals by a different control light that it has entered into a nonadaptation mode. Those indicator lights, or rather, hormones, may thus be regarded as markers of our adaptive reactions. They are relatively easy to measure, for instance in the blood or the urine, and can provide us with indirect information on the performance of the adaptive mechanisms in our brains. These parameters are, of course, more difficult to assess directly. The study of those beacons is akin to an attempt to understand the state of mind of unseen defenders in a besieged citadel by analyzing the more easily observable movements of supplies in the immediate vicinity.

These transitory hormonal changes are selective and hence generally beneficial in that they promote the restoration of the organism's equilibrium in relation to its environment. They are factors for adjusting to emergency situations, usually acute situations, and are more effective when short-lived. However, if it so happens that the behaviors they accompany do not attain their goal, they may become prolonged and go from being acute to chronic. Beyond a certain limit, the maintenance of a state of emergency perturbs normal physiological regulatory mechanisms. Exceptional, priority messages transmitted through the usual cell communication systems begin to interfere more and more with the coordination of organic functions and may indeed produce pathological conditions. Instead of being beneficial for our short-term ability to adapt, they are converted into vectors of somatization, that is to say, signals that help to change behavioral distress into organic disease. For example, norepinephrine, which can damage the cardiovascular system if hypersecreted, and also β-endorphin, which interferes with numerous functions and the immune system in particular, can act as relays between inappropriate behaviors and the derangement of certain physiological functions.

Based on this property, efforts have been made to use these hormonal markers as predictive indices of adaptation distress or, to put it another way, of the "biological cost" of strategies for responding to work rules and commands. Studies have been undertaken on the potential risk of morbidity (in this instance, occupational diseases) related to work schedules of bus drivers in the Netherlands or to a reorganization of work at the time when old London telephone switchboards were replaced by modern equipment. An excessively high number of adaptive episodes can have the same effects as frustrated behavior. It should be made clear, however, that these markers are only of statistical value. In view of the wide variations among individuals, they can be predictive of the risk for a population but not for a particular individual.

This is an important reservation. Following the work of the Swiss psychiatrist Eugen Bleuler, which now dates back a number of years, there had been high hopes for the development of a biological psychiatry that would have made it possible to fine-tune the diagnosis of psychiatric illnesses such as depression on the basis of "objective" biological and biochemical parameters. Legitimate though they were, these hopes came to naught. To oversimplify, (bearing in mind that problems of adaptation are only one factor), we can compare depression to the failure of an adaptive reaction that is gradually transformed into chronic maladjustment. In this sense, it does not, at least in its initial phase, have the characteristics of an organic disease; unlike diabetes or metabolic diseases, it cannot be explained by a dysfunction in the transmission or reception of messages due to a chemical lesion. It is more a matter of improper use of the emergency communication mode normally intended to cope with acute episodes of stress. This misuse of the state of emergency jams the functioning of the nerve cell communication systems. If the jamming continues, the organism tries to adapt to it by seeking a new equilibrium that differs from its usual equilibrium; it learns to live with the interference. However, it may then deviate too much from the adaptive behaviors that condition its ability to survive.

The mysterious efficacy of electroshock therapy is probably due to its ability to abruptly disrupt these alternate states of equilibrium. When the structure of complex combinations of signal exchanges is thus disrupted, we can imagine that they will reorganize along different lines, just as if we were to bet that a violent noise could help us restore the lines of a melody that our memory had distorted.

THE BRAIN AND IMMUNITY

Therefore, it is only when a short-range adjustment reaction persists too long that adaptation hormones turn into somatization vectors. They then contribute to increasing the probability of physiological disorders such as cardiac or immune deficiencies. Immune deficiencies are now the object of investigations in a new field known as neuroimmunology, which studies the interactions between the nervous and immune systems.

As stated earlier, nerve cells, glandular cells, and lymphocytes are able to express common signals; that is what we have referred to as the universality of communication systems. Some authors believe that this accounts for the functioning of somatization vectors. If nervous system mediators (neurotransmitters) can be recognized by a lymphocyte and, conversely, if a neuron can decipher immune response mediators, both types of cells can communicate directly and influence each other.

Such was the interpretation first given the discovery of the manner in which interleukin 1, an immune system mediator, is involved in the mechanism of fever. For all intents and purposes, it appeared as though the brain could be directly informed of the existence of an infectious state by this interleukin that is produced during activation of the immune system. It would respond to this signal by hyperthermia, a favorable climate for combating bacteria. The sharing of signals provided an explanation for some joint functional activities that are very important to the body. Parallel to this, it had been shown that activated lymphocytes were capable of secreting certain stress hormones. The conclusion was that during invasion by bacteria, the immune system could directly muster a state of emergency, bypassing the cerebral mechanisms normally in charge.

Unfortunately for this fine theory, the explanation now appears too simple. Signals have been found to be universal in all organs. Still, they do not circulate unrestrictedly throughout the body. It will be remembered that there are short-range and long-range signals. It happens that interleukins have a short range, and lymphocytes probably do not produce them in sufficient quantity to reach the brain easily—not to mention the fact that the brain is protected against flooding by long-distance signals by a chemical barrier, namely, the blood-brain barrier.

Recent data have now made it possible to clear up the reason for the misunderstanding: the nervous system itself produces interleukins that it uses as one of its own signals in temperature regulation. The same signals were simply selected independently by different organs. But beyond such independent selection, it was found that certain signals are used more frequently than others as part of a given physiological function. This finding is based on analyses of frequency conducted in the same spirit as those performed for the purposes of deciphering signs of some unknown writing. A comparison of the most frequent graphs with their context can provide clues as to their meaning or semantic function. In the case of cell communication, the meaning is of course given by the purpose of the function studied.

The first analyses of frequency relating to cell communication signals addressed three major body functions: energy balance, water metabolism, and reproduction. In each case, the number of exchanges involving a given signal was evaluated. In practice, this meant counting in all organs associated with a given function the types of cells capable of transmitting or recognizing the signal. For the reproductive function, for example, the neurons involved in sexual behavior were counted, as were the cells of sex glands and external genital organs. As could be anticipated by the theory of the universality of signals, all mediators studied (about 50) were involved in all functions, but very infrequently so (less than 5% of the types of cells analyzed). Yet, two more surprising observations were made. One family of signals (catecholamines and acetylcholine) appears to be involved at a high rate in all functions (more than half of all cell types analyzed). Conversely, other categories of molecules seem to be involved at a high rate in only a single function (somatostatin and glucagon in energy regulation, vasopressin and angiotensin in water metabolism, and finally LH-RH and a morphine-like peptide in reproductive functions).

These observations are still fragmentary and permit no real conclusion as to the "semantics" of cell communication or the rules that determine its meaning. But they do show that far from being a random procedure, the selection of signals follows precise laws that we are barely beginning to decipher correctly and which apply to all physiological functions. In this sense, the theory of cell communication—the body of universal rules common to all organs and all organisms—is radically changing the traditional views of the physiology of functions or organs.

Let us return to the local, short-range signals. What purpose can they serve if they do not provide the means of communication between distant organs? We have already referred to their role in nearby communication as synaptic transmitters, for example, or as autocrine amplifiers of cell activity. However, circulating cells such as blood cells can also provide home delivery of their message by autotransportation to a target tissue. Lymphocytes can temporarily take charge of communications in an organ they colonize. As we know, localized infections are followed by an "invasion," in fact a colonization of the sick organ by immune system cells which are thus able to bypass the signals of the infected tissue to carry out their "clean-up" operations.

In these conditions, how are relatively distant organs like the brain and organs of the immune system able to communicate with each other? They do so essentially by resorting to long-distance signals, such as pituitary hormones, and to "somatization vectors" called up in emergency situations. The so-called stress hormones—namely β-endorphin, adrenocorticotropic hormone and adrenal hormones—are often used for this purpose, but so are some others (growth hormones, sex hormones) that are involved in the body's adaptation to the environment and possess immunomodulating effects. In acute situations, these effects do not appear to be of any particular consequence. But when they persist, in situations of chronic failure of adaptation, they can interfere with the immune defenses. As we acquire a better understanding of these immunomodulation phenomena, we will no doubt find a scientific explanation for observations that have long been acknowledged on the basis of common sense and have been confirmed by clinical studies: for example, that the degree of resistance to tumoral processes or the remission of certain cancers can depend on the patient's adaptive strategy, that is to say, the patient's own reaction to the disease.

IV

THE EVOLUTION OF BIOLOGICAL COMMUNICATION SYSTEMS

The evolution of the systems of communication is a fascinating area of research: it seeks to determine how the extraordinarily complex cell-signaling mechanisms were able to develop from relatively elementary signals exchanged by primitive organisms with their environment. Here, in a new form, we are again confronted by the old question of the relation between simplicity and complexity in biology.

The theory of cell communication has uncovered rules that have remained more or less unchanged during the course of evolution. From these rules, we can imagine scenarios and propose hypotheses about the stages of development of increasingly complex systems. Those scenarios are still highly speculative but will perhaps open up new avenues of research.

The elementary communication systems of simple organisms—say, a bacterium or paramecium directly immersed in a liquid medium—fulfill the need for securing information regarding their environment:

what are the available nutrients, how has the cell already transformed them? In connection with the applications of the law of mass action, we have emphasized the analogy of the exchanges of signals with enzymatic reactions. A cell comes in contact with some raw material and converts it by means of enzymes. Those enzymes already amount to a system of specific recognition through their particular affinity for given substrates. Therefore, the passage of an enzyme-substrate coupling to receptor-signal coupling represents a progressive transformation from an enzymatic system to a receptor system. In the course of this evolution, the organization of the receptor itself has grown in complexity. Very ancient organisms such as yeasts already have amino acid sequences that strangely resemble the different components of our receptors: recognition sequences, G proteins, and phosphorylating enzymes. However, in primitive life forms, those sequences are often coded by the same gene, and then translated from end to end in the same protein. It appears as though the components of an elaborate communication system were already present in this archetypal receptor expressed by the yeast and gradually became independent of each other in the course of evolution. The different families of molecules involved in recognition, coupling, phosphorylation, and ion transport were eventually coded by different genes as a result of rearrangements and of recombinations of their DNA sequences.

The advantage of such rearrangements is clear. By cutting up the original molecule into segments for different functions and then promoting diversification of the separate segments, the number of possible coupling combinations is increased considerably. The process is somewhat reminiscent of a Lego set in which the shape of certain types of blocks has been slightly altered; after reassembling them, infinite variations from the original pattern could be produced. In the case of the coupling mechanisms we are discussing, the significance of this metaphor is that such a modular system makes it possible to introduce great variety into the relation between the different types of receptors and intracellular effectors.

However, the genes coding for each component are themselves diversified, resulting in the development of "families" of homologous molecules. A minor mutation, affecting for instance a single nucleotide of a gene comprising thousands of them, can modify the properties of the protein it encodes. If these changes alter the function of the molecule too much, a reverse selection will take place. Sometimes, these modifications will have a neutral effect, with the old and new molecules coexisting. But slight changes in the recognition sequence or in a domain of the molecule determining its anchorage to the cell membrane may result in a new property. In this way, families of molecules are formed that have similar properties but differ slightly in their affinity for a signal or in their enzymatic specificity.

Such methods of diversification are found at all levels in the case of receptors as well as signals and coupling molecules. The great family of opioids (opiate-like substances), for example, comprises numerous signals that all have the same general properties but possess preferential affinities for subclasses of receptors, which themselves are diversified. Certain signals in this family, such as met-enkephalin and leu-enkephalin, differ by only a single amino acid.

Gene divergence or duplication can result in the coding for different signals by the same gene or, just the opposite, in the coding for identical or similar signals by different genes. The same transformations have affected the famous coupling proteins called G proteins, but also the ion channels that possess great structural homologies, as well as some differences that account for their specificity for a given ion. The same process holds in molecules that take part in the intracellular transport of the cell products: slightly modified patterns "personalize" the addressing procedures of each cell. Thus, the evolution of signaling molecules appears to follow simple rules. Some classes of information molecules variously responsible for recognition, coupling, ion exchange, or phosphorylation, and which themselves derive from a common ancestor, have become diversified, each for its own account, and this has enriched considerably the cell's communication repertoire. From an elementary membrane protein like that of yeast, mentioned earlier, some simple rules of diversification have produced an extraordinarily complex communication system.

How does the cell find its way through such complexity? This is where the shunting function of G proteins as intermediaries comes in. Through them, the vast numbers of signals are channeled in the direction of a few effectors which, through mediation of the movement of ions or by phosphorylations, translate them into simpler messages. In this way, the coupling mechanisms act as simplifiers; they help to make sense out of codes.

This diversification of families of signals and receptors also poses another problem: how to preserve intelligible signaling despite the growing complexity of its elements. In other words, how can the theoretically independent signal and receptor diversification mechanisms produce compatible keys and locks?

To try to answer this question, we can follow two approaches. The first is based on observations suggesting that redundant exchanges of signals are likely to be found where the function they serve is vital for the organism. Of course, such redundancy limits the consequences that could result from the loss of compatibility of a signal-receptor complex. The second is based on the still problematic hypothesis of an active selection mechanism for the compatibility of keys and locks.

It appears that the more selective and important to the survival of an individual or a species a function is, the more its communication systems are redundant. The cells and organs associated synergistically in accomplishing a given function are linked together by a number of signals repeating the same messages, thus by "synonymous" signals, so to speak. For example, the pituitary cell that secretes growth hormone can be stimulated by at least six molecules produced by neurons of the brain-hormone interface; it also receives an inhibitory signal, somatostatin. One of the activating signals also seems capable of inhibiting somatostatin. We therefore have at the same time a redundancy of synonymous signals (stimulatory) and a redundancy of the effect of a signal that can simultaneously stimulate the cell directly and repress its inhibitor.

Paradoxically, we note that the greater the redundancy, the greater is the ability of the system to fluctuate within limits compatible with the function. Thus, the redundant system provides a multiple security mechanism against the temporary failure of a signal. This is no doubt the reason why pathological conditions linked to signal-exchange abnormalities are rare outside of adverse genetic modifications of signaling molecules. The rules for the selection of redundancy are still poorly known; the signal frequency analyses we discussed previously should help us to understand them better.

The hypothesis of a concerted selection of compatible signals is based on research in the area of ontogenic development, in other words on the development of organs and organisms. Of course, organogenesis involves cell multiplication. But multiplied cells still undergo several transformations and several migrations before finding their permanent places in the organs to which they ultimately belong. Their positioning follows precise rules: the marking of migration pathways, notably by the cell adhesion molecules (CAMs) discussed above; the influence of neighboring cells that help determine the modality and timetable for definitive differentiation. When identical embryonic nerve cells are transplanted to two different structures of the central nervous system, and thus to environments that have different signals, the transplanted cells modulate the expression of their genome (gene pool) according to the surrounding cells. In this case, the signals coming from adult cells that are characteristic of the cerebral structure where they are located can influence the choice of signals that the transplanted cell will use.

Such influences are of particular importance in the nervous system because adult neurons, unlike most other cells, have lost the ability to divide. As a result, the connections they establish with one another are reversible, which may explain why certain embryonic neuronal junctures are transitory in nature. It has been observed that a large number of connections, for example in the cortex, do not manage to stabilize at the time that the nervous system is developing and disappear after a few days. We still do not know why. It may be that these temporary connections, which might be compared to trial marriages, do not ultimately produce compatible communication modalities. But it is also possible that their purpose is to allow two cells temporarily coming into contact to exchange information that might subsequently be useful throughout their lives, without requiring a permanent connection to be established between them. Whatever this purpose may be, the other connections—the ones that do not degenerate—are stabilized, form permanent synapses, and establish a compatible communication system.

SOME DEGREE OF INDETERMINATENESS

The growth of neurons, like the stabilization of temporary connections, follows precise signaling rules strictly inscribed in their genomes. The neurites—nerve cell processes (axons and dendrites) that grow as the neuron develops—follow pathways that are perfectly marked out. Other signals, similar to the cell adhesion molecules discussed above, stop their migration and put them in contact with a target neuron.

At this stage, any recognition error will result in severe developmental defects. Malformations involving the "wiring" of neurons in the cerebellum, for instance, have been studied by several French groups in mutant mice that have lost a particular category of target neuron. When they reach the vicinity of the missing target, the growing fibers are unable to become differentiated; having no stop signal, they continue on their course. In most cases, such mutations are lethal, meaning that they are incompatible with survival beyond a certain stage of development. However, some degree of freedom is left in this developmental process that is almost entirely predestined according to the laws of genetics. Recognition mechanisms are specific for cell types and not for individual neurons, so that neuronal connection is not predetermined cell by cell but over-all, population by population. Therefore, there remains in the system a certain margin of indeterminateness (neural plasticity), so that a neurite belonging to a given population can theoretically make connections with any neuron in its target population.

Since several different neurites can synapse with the same neuron, different populations of fibers may compete for the same target. Given such conditions, any factor liable to affect the growth of a category of neurites can modify the conditions of this competition and thus favor one type of synapse over another. These factors are often hormones, estradiol and progesterone, which were mentioned earlier, or else adrenal or thyroid hormones, which have astonishing morphogenetic properties: in the early stages of development, they can modify the neurons of certain cerebral structures irreversibly. Even in adults, they are still capable of altering, albeit temporarily, the shape or connection of certain neurons.

External factors can also affect the "wiring" of the brain, for instance the conditions in which the systems are activated at the time they are organizing (generally during fetal development, though certain synapses do not mature until after birth). This explains the interference of epigenetic (circumstantial or accidental) factors—linked for example to the physiology of the mother or, more generally, to the environment of the fetus or neonate—with the strict genetic rules for the development of the nervous system. A case of such epigenetic interferences that has been well documented relates to the differentiation of the first synaptic relay of the visual projections of the retina. Experiments on kittens have shown that if the visual space is deformed—for instance by allowing only vertical planes to appear—the synapses involved do not develop according to their normal plan during the critical phase when this nervous pathway is first put into operation. Thus, the organization of a nervous structure can be modified by the animal's first contact with its perceptive environment; once it has stabilized, this change is irreversible. In the context of this experiment, the kitten derives no benefit from its new development plan. But the fact that the very structure of the sensory information-processing networks can adapt to lasting changes in the perceptive universe no doubt offers selective advantages under given conditions or for certain animal populations.

Temperature, the quantity of light, or the salt concentration of the water for aquatic species, all appear to be factors able to influence the development of the nervous system. The relative plasticity (indeterminateness) of neuron-to-neuron contacts renders the question of their compatibility even more acute. If neurons are able, even to a slight degree, to choose their partners, it becomes of the utmost importance for them to have means of concerted action to express compatible keys and locks.

There is a molecular basis for this concerted action, although for the time being we can only theorize in this respect. One hypothesis has aroused interest; it was formulated in 1984 by Ed Blalock, an American researcher. The data are still fragmentary, and his theory has given rise to lively controversies. It implies that at the time of their initial contact, the two cell partners involved "reach an agreement" on the common expression of genes encoding complementary molecules. Blalock suggested that the information coding for a signal and its receptor might be contained in two complementary DNA strands in the same gene. We know that genes consist of two complementary strands; only one is presented in a direction permitting transcription, while the other is the negative of the first one, so to speak. The "antisense" strand, which is essential for gene replication, also plays an important role in repair, precisely because it serves as a negative template. In principle, it should be noncoding. In the theory proposed by Blalock, the signal sequence is coded for by the strand normally transcribed, but the complementary strand contains the information coding for the receptor. Thus, two cells could reach agreement on a compatible communication system, and hence on the expression of a signal and its specific receptor, simply by activating the same gene transcribed by one cell and by the other in the reverse direction.

However, this hypothesis has run into several theoretical objections and for the time being can only be considered as a research lead.

Be that as it may, the development plan of living organisms can gain a few degrees of freedom by acting in different ways on the following three principal constraints: the rigid genetic rules that favor the invariance of the species, the rules of integration of epigenetic factors that make the species more adaptable to changes in the environment, and lastly the rules for concerted action that can insure the compatibility of the communication systems. This complex interplay enables the organisms to escape from the extreme stereotypy that the rigid application of the genetic rules alone would imply: without this small amount of room to maneuver, the adaptability of individuals might well be sacrificed to the mechanisms that insure the invariance of the species. It is probably the lack of such developmental flexibility (plasticity) that gives invertebrates their stereotyped behavior.

One point, however, should be stressed: the degrees of freedom that the developing organism enjoys apply only marginally, individual by individual; they cannot be transmitted to its descendants. In the final analysis, the genetic rules of embryogenesis have the last word, and these rules themselves, operate on several levels, arranged in hierarchical order. First of all, there is a segment specification: special genes known as homeotic genes determine a general organization plan for the body's major subdivisions and the position of their boundaries—for example, the segments of insects, or the head, thorax and limbs of vertebrates. Next comes tissue specification, to provide internal coherence to the organs; this is the area of operation of another large family of genes, the oncogenes, described in 1983 by Dominique Stehelin, a researcher from Lille, France, who at the time was working with Michael Bishop and Harold Varmus in San Francisco. This discovery won them the Nobel prize. Like homeotic genes, oncogenes are normally expressed only at definite stages of embryonic development. For the harmonious implementation of the architectural design of each tissue, the specification of the orderly sequence of events for the formation and development of organs has to precede the specification of their regulation, and hence of their communication system. No doubt, this constraint is not unrelated to the fact that several of these oncogenes code for proteins closely related to information molecules and particularly those of the G protein family. Actually, some oncogenes have the ability to replace the cell's locks or coupling components and to bypass its control mechanisms until it finds its permanent place in the body's architecture.

The transient predominance of the specifications for development over the ones for function protects the organs against inopportune signals that could interfere with their growth. But this system of protection may turn against the organism: the inappropriate expression of the oncogenes can lead to catastrophe and cause tumors—hence the name "tumor genes." Their cancer-causing potential is due to what is really their strong point during embryogenesis: the ability to partially shield the cell from signals in its environment.

However, these hierarchical levels of genetic rules are not sufficient to explain the evolution of the species, namely the whole set of processes by which organisms became diversified and increasingly complex, progressively improving their repertoire of communication and adaptive performances. Is the sum total of chance mutations really sufficient to explain the surprising continuity of the changes which, on a geological time scale, have freed the living world from the servitude of a hostile and unpredictable environment and the strategies that led to the victory of a negentropic order, that is to say, one opposed to physical laws that rather tend to break up the order? There is a contradiction of long standing between the intuition that evolution follows a logical plan and scientific evidence that this plan is not provided in advance, that it can only be deciphered after the fact, based on what the selection cutter has left us, preserving only the most adaptive innovations. Some as yet tenuous clues may foreshadow a reconciliation of these antinomic principles.

STRATEGIES FOR EXPLORING DIVERSITY

Also to receive the Nobel prize, in 1987, was the Japanese-American immunologist Susumu Tonegawa for elucidating one of the principal enigmas in his field. Antibody-producing cells possess the ability to find an effective parry against invasion by foreign substances. However, the variety of nonself materials that can invade the body is potentially infinite; the nature of the invasion is therefore wholly unpredictable. How can elements of the genetic code that have been selected on the basis of predictable scenarios provide appropriate responses for an infinite diversity of challenges? Tonegawa showed that some of the cells of the immune system subjected to a very strong selection pressure—like the invasion by nonself molecules during a microbial infection—were capable of setting in motion an active diversification process of their genomic expression. The process involves the excision of gene segments followed by recombinations through alternative splicing of the excised segments. Certain parts may also be suppressed, inverted, or modified before they are reinserted in the gene. In fact, all these reactions result in the production of new genes.

These changes, however, do not take place at random. They concern only specifically located segments in the genes called variable or hypervariable regions and do not affect others. In the variable domains, recombinations and restructuring do not occur randomly but follow precise rules (for example, the excision or splicing takes place only at given nucleotide sequences, since some sequences are more easily inverted than others). This body of rules determines the strategies for actively exploring diversity. When implemented simultaneously by millions of cells, this strategy is truly innovative; it has a reasonable probability of culminating in creating an effective parry. The first cell to find the right defense is so informed when a connection is established between the foreign antigen (the key) and its new antibody (the lock). It is immediately selected from among all the others which then give up the search, and the proliferation of identical cells ensues, thus setting up an effective barrier against the aggressor. This directed mutation is hereditary for only a single cell type; it is a somatic mutation that will not be inscribed in the genome of the other tissues and will not be transmitted to the germ cells.

While exercising the caution required for any reasoning based on analogy, we might try to extrapolate this process to evolution. Could it be that a very strong selection pressure resulting from some change in the environment dramatic enough to threaten the very survival of a species led to a similar exploration strategy for genetic diversity? Such a scheme cannot be easily verified; somatic mutations of immune cells occur on a time scale of hours, whereas mutations related to evolution require millions of years. Yet such a teleonomic hypothesis could provide scientific backing for our feeling that evolution proceeds in a straight line. It may perhaps help us to better understand how the adaptability of species and their ability to escape stereotypes increase as we ascend the evolutionary scale.

V

SCIENCE: FROM EPISTEMOLOGY TO POLICY

As new signals are discovered and their functions are being deciphered, the as yet tentative outline of a general theory of cell communication is beginning to emerge. No doubt, the semantics of this communication is still a mystery. For the time being, we are trying to make an inventory of these signs, to list their most common combinations and to match the message they compose with a meaning—in this context with a biological function. As of now, we are still collecting data to try to understand the syntax of the cell's language.

But already we can see the emergence of rules common to the regulation of all functions. Tomorrow's physiology will no longer look like classical physiology, with its boundaries between disciplines within which the function of different organs—including the immune or cardiovascular systems, the kidneys or the brain—was the object of separate studies conducted by different groups that rarely communicated with each other. It will rather take the form of a general theory of regulations, as a "transversal" discipline, so to speak, that will cut across all of them and tend to replace the comparison of the unique features of organs by a comparison of particular uses of common rules. In this sense, its development is like that of histology—mentioned by Dominique Lecourt in his introduction—which made the cell the structural unit, the "atom" common to all organs.

With this prospect, we can foresee the emergence of a new set of axioms, that is to say, a coherent set of theories and explanatory schemas. Other sciences, such as physics or chemistry, have long since moved toward the development of a set of axioms; technological advances coupled with increasingly sophisticated methods for deciphering complexity will no doubt make it possible to extend this development to all scientific fields.

This opening up of borders is part of a much more general movement in modern science. Starting from a global approach to the world, which encompassed the study of the physical world as well as mathematics and philosophy, science gradually became partitioned into large sectors which in turn branched out into disciplines. As a result, scientific communities, learned societies, and specialized scientific journals became more and more dispersed. Today, we are beginning to see a trend in the opposite direction: associations are now collaborating or merging with each other, journals addressed to the general reader are growing in importance in contrast to journals for the specialist, and scientists are trying to relativize their special know-how in favor of thinking in global terms.

So far as biology is concerned, the cell communication theory is its second body of axioms, after molecular genetics. With a renovated representation of physiological processes, we will gradually move from the traditional approach to life sciences—inducing rules on the basis of experience—to an approach where facts are compared with theory. This development will in turn make it possible to deduce an increasing number of applications—medical, pharmaceutical, and industrial—from theory, whereas today the applications of biology are still often empirical.

A more axiomatic context will also transform the way we interpret pathology and make us look for the initial causes, somctimes global, of impaired health. With less fragmented objects of scientific study—organs or functions—and a common language to describe them, the major medical problems will help put into perspective the data accumulated by biological science. It will be possible to mobilize all the resources of theory to answer the challenges of disease in configurations that will be based more on the nature of particular pathological problems than on splitting them up into medical specialities.

Other problematics, other large questions—the sciences of evolution, the study of the major biological mechanisms for maintaining a state of equilibrium in relation to the environment—will also suggest new structuring principles to biological science and will benefit in return from its latest research strategies.

In this framework, we can anticipate the development of novel therapies. The administration of medicinal drugs still consists in replacing a defective signal by an artificial molecule that mimics its effect. But of course the purpose of the exercise is not to replace a signal; it is to restore regulation and hence to substitute the proper message for a truncated one. We will probably turn to replacement therapies based on the simultaneous or sequential administration of several signals, the ones that determine the meaning of the message.

We have just seen how the movement to decompartmentalize and open closed doors should have an impact on the practice of biology and the strategy of its applications. However, we need to adjust our training and research policies to be able to take advantage of this development.

The necessary adjustments will confront us with three challenges. The first has to do with the cultural image of biology, or the way that nonspecialists, ordinary citizens, see the life sciences. Our educational programs have integrated the image of physics much better than that of biology. When the mass education plan for primary and secondary schools was established at the end of the last century, physics was a much more apt model for demonstrating the scientific method than were the other experimental sciences. The manner in which it already combined application, experimentation, and theory and how it sealed this alliance by the use of mathematical formulas was an exemplary model, at least in principle, for the comparison of hypotheses with observable reality. Moreover, physics had a definite pedagogic value for making the research processes tangible.

Our entire educational system has been steeped in this approach. Even those who followed very different career paths have retained from this experience some reference points, an unanticipated familiarity. From its original mold as a descriptive discipline, biology as it is taught in elementary schools has retained somewhat of an "object lesson" style based on observation of animals and the mechanics of their function. This mechanistic representation of the living world actually bears the mark of an archaic cultural context, that of the late 19th century; it has aged more than that of the world of physics.

The metaphors applied to physiology at that time nearly always referred to machines. Some fifty years later, this mechanistic vision was tempered by electrical metaphors, with the brain being conceived as a very powerful telephone system. And, of course, the metaphors used in the writings of today's biologists come from computer science, with the "software systems" of cerebral communication programming the "hardware" of the brain, in this instance the "wiring" of the nervous system.

No metaphor is really explanatory; rather, it reflects the cultural references through which we have been conditioned to decipher reality. But these cultural references play an important role in the way we look at the world. We have not yet really learned how to teach biology through the cultural references of our time, which perhaps explains why for many of our contemporaries its picture is still somewhat blurred.

The second challenge is addressed to how our system of higher education is structured. Universities and institutes of higher learning—in which the life sciences are not really an established or even accepted part of the general curriculum—have not yet sized up the importance of decompartmentalization in this discipline. Teaching is generally in the mold of the old courses by individual subjects—anatomy, physiology, endocrinology, or immunology—which impedes the widespread adoption of the new "transversal" disciplines that cut across subject lines. Cell biology and molecular genetics are still often taught separately, while they ought to contribute to the restructuring of the traditional fields.

What applies to separation into subjects also holds true for career training. Clinical research, pharmaceutical innovation, and thus a good part of medical progress, will depend increasingly on mastering biological axiomatics that are mostly not taught in engineering schools, schools of medicine or pharmacy, and are very unevenly taught in schools of science. Since there is no intercommunication in career training and each branch has a virtual monopoly on employment prospects, many activity sectors have not benefitted from decompartmentalization of disciplines. This represents a great handicap for medical research as well as for industry.

We in France have also failed to adopt one of the "master trump cards" of American research, which ensures that new biological concepts and tools are transmitted rapidly to all those engaged in a particular field, in the universities as well as in industry, namely a coherent postdoctoral system. It is at this level of education, which precedes a permanent career choice, that there is the best mix of disciplines, the interpenetration of basic and applied approaches, the establishment of international solidarity. Of course, even if the American model of research and development is very effective, it does have its faults; and even its qualities cannot always be transposed to Europe. But setting up a postdoctoral program on a European scale is an urgent necessity for the life sciences. Curiously, if we look at American science on this score, we see that postdoctoral practices are not the same in all experimental sciences; they seem to be most fertile in biology, chemistry, and the earth sciences; in physics, it is generally through other modalities that this essential function of accomplishing a rapid mix of skills is ensured.

The third challenge concerns the organization of our research potential: How are we to enable it to take better advantage of recent developments while becoming more responsive to the demands of society, namely what the community can expect from biology in economic terms or in terms of health? In other words, how can research be managed to take the new orientation of life sciences into account?

Opening up the disciplinary borders allows specialists formerly unaware of each other to acquire a common language. During a recent "transversal" symposium attended by researchers in different branches, an immunologist was listening to the paper of a neurologist: "It's incredible," he complained, "I was about to say exactly the same thing in a little while . . . ." Even though the two researchers worked in the same city, they did not know each other. But the logic of scientific decompartmentalization had led them to the same conclusions, starting from different models and quite independently of one another.

Through common languages, openness naturally also facilitates cooperation between different specialties and accelerates the emergence of new ideas, each participant enriching joint strategies through contributions reflecting his or her own intellectual approach.

Here, the prospects for decompartmentalization and the demands of society intersect.

Societal needs are expressed in terms of concrete objectives that do not necessarily coincide with the logic adopted by each discipline. The objective to be reached does not always require the setting up of a permanent structure or team. It is often by defining a research operation limited in time, like a sort of commando operation, that the answer will be found most effectively. To take the example of AIDS, we know that the HIV virus affects certain cells through an effective use of their recognition mechanisms. In a way, it becomes subservient to the addressing and signaling properties of the tissues. In order to investigate the modality of infection of a particular tissue, say, nervous tissue, it would not be very smart to try converting an increasing number of neurobiologists to virologists—which might be the initial temptation of some well-intentioned technocrat!—even if, for historical reasons related to the long time lapse since the major epidemics, we are now terribly short of virologists (at the same time that we have a large number of neurobiologists). You do not necessarily get good football or rugby players by trying to convert soccer players!

Instead of pursuing artificial reconversions, it would be better to promote the creation of "commandos" comprising the complementary skills required for each mission. In such units, we can assume that a greater number of scientists will participate in collective endeavors—whose "public" usefulness must of course be evaluated, as must all research activity—without losing their autonomy, a factor indispensable to the practice of their profession.

However, certain operations calling for very diversified contributions must be long-term undertakings, for example when there is a broader and more general objective: understanding major pathologies such as cancer, cardiovascular disease, developmental or geriatric diseases, or making new biological tools available to paleontology and the evolutionary sciences, or for analyzing the great states of biological equilibrium within the environment.

For such operations to be successful, we will have to learn how to manage research by changing some of our habits. For example, we will have to institute on-site policies to assemble specialists from different disciplines to work on long-range projects without letting their specificity slip away. In plain language, we will have to learn how to assemble at well chosen sites the many pieces required for mounting a long-term operation, carefully selecting all the partners who will participate. On-site operations will become increasingly important as part of an overall national development policy.

Thus defined, these operations will gradually provide the framework for departments on the American model, with one level for strategic studies (where the best possible proportioning of skills will be determined and where the long-term goals will be worked out in consultation with all the participants), and another level for research where the freedom of action of the research groups will be jealously guarded. This last point appears to be essential: research, by its very nature, remains unpredictable because it attempts to decipher the unknown; the skills it requires change rapidly. Therefore, it cannot abide rigid hierarchies and unyielding structures, but of course it must not yield to the opposite temptation of anarchy. The keys to our scientific future lie in a harmonious development of the hierarchies of power and the hierarchies of skills, in learning to determine the optimum level of autonomy and responsibility for the research groups.

The challenges that the life sciences present to the political powers that be include a better adjustment to the demands of society, refurbishing of the cultural image of biology, renovation of career training and opportunities, national planning and development. Will they size up the full measure of the situation? And above all—even if the cost sometimes appears too high, if the result appears to come late, and the privilege seems exorbitant in terms of the customary administrative practices—will they always be able to respect this essential condition of any innovative activity: creative freedom?

BIBLIOGRAPHY

DANTZER, R., *L'Illusion psychosomatique* [The psychosomatic illusion],* Odile Jacob, Paris, 1989.

GILMAN, A.G., "G Proteins: Transducers of Receptor-Generated Signals," in *Annual Reviews of Biochemistry* **56**:615-649, 1987.

HESCH, R.U., *Endokrinologie* [Endocrinology], in the series *Innere Medizin der Gegenwart* [Internal medicine of today],* 2 vol., Urban und Schwarzenberg, Munich, 1989.

HILLE, B., *Ionic Channels in Excitable Membranes*, Sinamer Associates, Inc., Sunderland, Massachusetts, 1984.

KACZMAREK, C.K. and LEVITAN, I.B., *Neuromodulation*, Oxford University Press, 1987.

KORDON, C. and DEGOS, L., *Communication cellulaire et pathologie* [Cell communication and pathology],* INSERM/John Libbey Eurotext, Paris, 1988.

ROITT, BROSTOFF, and MALE, *Immunology*, Gower Medical Publishing House, London, 1985.

STRIYER, L. and BOURNE, H.R.: "G Proteins: A Family of Signal Transducers" in *Annual Reviews of Cell Biology* **2**:391-419, 1986.

VINCENT, J.D., *Biologie des passions* [The biology of passions],* Odile Jacob, Paris, 1986.

WADE, N., *La Course au Nobel* [The Nobel race],* Sylvie Messinger, Paris, 1981.

* These references have not been published in English.